CREATURE FEATURES: CLASSIFY ANIMALS!

IT'S A FISH!

BY NATALIE HUMPHREY

Please visit our website, www.garethstevens.com. For a free color catalog of all our high-quality books, call toll free 1-800-542-2595 or fax 1-877-542-2596.

Cataloging-in-Publication Data
Names: Humphrey, Natalie.
Title: It's a fish! / Natalie Humphrey.
Description: Buffalo, NY : Gareth Stevens Publishing, 2025. | Series: Creature features: classify animals! | Includes glossary and index.
Identifiers: ISBN 9781482466836 (pbk.) | ISBN 9781482466843 (library bound) | ISBN 9781482466850 (ebook)
Subjects: LCSH: Fishes–Juvenile literature.
Classification: LCC QL617.2 H848 2025 | DDC 597–dc23

Published in 2025 by
Gareth Stevens Publishing
2544 Clinton Street
Buffalo, NY 14224

Designer: Andrea Davison-Bartolotta
Editor: Natalie Humphrey

Photo credits: Cover Jsegalexplore/Shutterstock.com; series art (backgrounds) Honyojima/Shutterstock.com; p. 5 (bottom) Lotus_studio/Shutterstock.com; p. 5 (top) Rich Carey/Shutterstock.com; p. 7 Worraket/Shutterstock.com; p. 9 Nosyrevy/Shutterstock.com; p. 11 (inset) BlueRingMedia/Shutterstock.com; p. 11 (main) Inosensius Julio Tolo/Shutterstock.com; p. 13 (bottom) makolovesfish/Shutterstock.com; p. 13 (top) unterwegs/Shutterstock.com; p. 15 Wirestock Creators/Shutterstock.com; p. 17 (bottom) O partime photo/Shutterstock.com; p. 17 (top) Studio 37/Shutterstock.com; p. 19 (top) Rostislav Stefanek/Shutterstock.com; p. 19 (bottom) SVITO-Time/Shutterstock.com; p. 21 TAUFIK ART/Shutterstock.com.

Printed in the United States of America

CPSIA compliance information: Batch #CS25GS: For further information contact Gareth Stevens, New York, New York at 1-800-542-2595.

CONTENTS

Boldface words appear in the glossary.

Is That a Fish?

What is a **cold-blooded** animal that spends all its life in the water? Fish! There are more than 30,000 **species** of fish that scientists know about, and new fish are discovered every year. At least 11 fish were discovered in 2022 alone!

What Makes a Fish?

All fish have at least one fin. They have **gills**. Fish are covered in **scales** or skin. But they can look quite different from one another! Fish can be bigger than a school bus or so small they're hard to spot. They can be many colors or all gray.

FINS
GILLS
SCALES

Fish Fins

Fish use their fins to move through the water. Fins may look different from species to species. Many fish have around five fins. Some fish, including most eels, only have one fin. Sharks, such as hammerhead sharks, usually have eight fins!

A Fish's Gills

A fish's gills are found on both sides of its head, right behind the **jaw**. When a fish needs to breathe, it opens its mouth to bring in water. The water flows over its gills, and the gills pull **oxygen** out of the water.

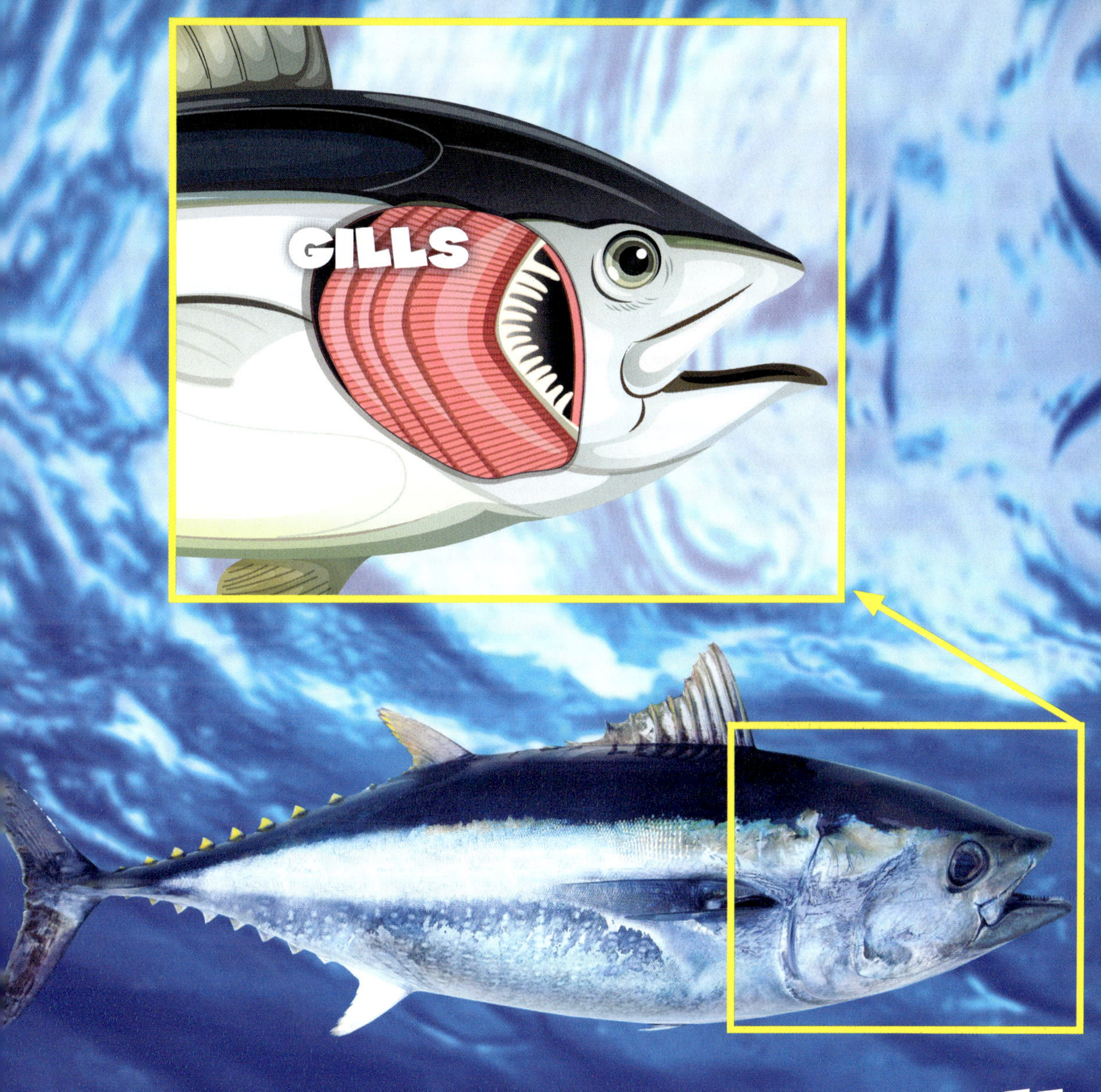
GILLS

Life in an Egg

Most fish start their lives in an egg. Some fish lay hundreds of eggs into the water. They may hide their eggs in plants. Some fish keep their eggs in nests made of sand, stones, or even bubbles. Some fish hold their eggs in their mouth!

EGGS

The eggs of some fish, including some kinds of rockfish, stay inside the mother. The eggs **hatch** inside of her, and the babies are born live. Some kinds of sharks and rays have live babies too. Baby fish are commonly called fry.

ROCKFISH

Growing Up

Baby fish aren't often cared for by their parents as they grow. Most fish grow for their whole lives. Smaller fish usually become adults in a few weeks or months. Some larger fish, including great white sharks, can take more than 26 years!

Around the World

Fish can be found living in waterways around the world. Some fish, such as tuna and swordfish, live in the salty water of the oceans. Other fish, like walleye or largemouth bass, live in fresh water found in rivers, lakes, and streams. Some fish can live in either!

LARGEMOUTH BASS
SWORDFISH

Helping Fish

Pollution and overfishing are some of the biggest dangers to fish. Overfishing is when people catch too much of a kind of fish in one spot and the number of fish there becomes low. By caring for Earth and learning more about fish, we can help them continue to live around the world!

THE OCEAN®
CLEANUP

GLOSSARY

cold-blooded: Having a body temperature that's the same as the temperature of the surroundings.

gill: The body part that fish use to breathe in water.

hatch: To break out of an egg.

jaw: One of the two bones of the face where teeth grow.

oxygen: A colorless, odorless gas that many animals, including people, need to breathe.

pollution: The act of making water or land unsafe for people or animals by depositing harmful matter in it.

scale: One of the small, flat plates that cover a fish or reptile's body.

species: A group of plants or animals that are all of the same kind.

FOR MORE INFORMATION

BOOKS

Jaycox, Jaclyn. *Unusual Life Cycles of Fish.* North Mankato, MN: Capstone Press, 2022.

Parker, Steve. *Fish.* New York, NY: DK, 2022.

WEBSITES

Britannica Kids: Fish
kids.britannica.com/kids/article/fish/353130
Discover more facts about what makes a fish a fish.

National Geographic Kids: Fish
kids.nationalgeographic.com/animals/fish
Learn about the different kinds of fish found around the world.

Publisher's note to educators and parents: Our editors have carefully reviewed these websites to ensure that they are suitable for students. Many websites change frequently, however, and we cannot guarantee that a site's future contents will continue to meet our high standards of quality and educational value. Be advised that students should be closely supervised whenever they access the internet.

INDEX